Igeddit Mathworks™

Color Coded Step by Step Instructions

The
Addition
Subtraction
Multiplication
and
Division
of
Fractions

Wanda Y. Brooks

Igeddit Creative Concepts, LLC.

Igeddit Mathworks™

Color Coded Step by Step Instructions

Author:	Wanda Brooks
Editor:	Jacqueline Brooks
Cover Design:	John Marrero, Wanda Brooks
Content Graphics:	Wanda Brooks

Igeddit Mathworks™
Addition, Subtraction, Multiplication and Division of fractions
Table of Contents

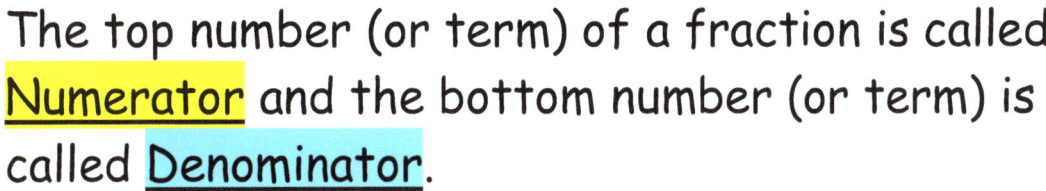

Retro-Lesson A

The top number (or term) of a fraction is called <mark>Numerator</mark> and the bottom number (or term) is called <mark>Denominator</mark>.

$$\frac{2}{5} = \frac{numerator\ is\ 2}{denominator\ is\ 5}$$

Proper Fraction

The <mark>numerator</mark> is <u>smaller than</u> the <mark>denominator</mark>.

$$\frac{4}{7} \Rightarrow 4 < 7 \Rightarrow proper\ fraction$$

Improper Fraction

The <mark>numerator</mark> is <u>larger than</u> the <mark>denominator</mark>.

$$\frac{8}{5} \Rightarrow 8 > 5 \Rightarrow improper\ fraction$$

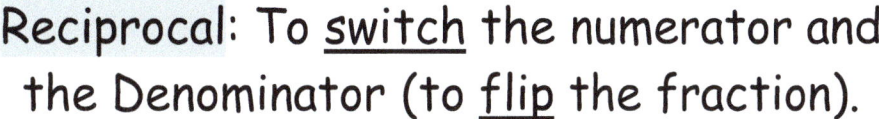

Retro-Lesson B

Reciprocal: To <u>switch</u> the numerator and the Denominator (to <u>flip</u> the fraction).

$$\text{the reciprocal of } \frac{2}{3} \text{ is } \frac{3}{2}$$

Every fraction with Matching Numerator and Denominator Equals One.

$$\frac{2}{2} = 1, \quad \frac{123}{123} = 1, \quad \frac{5x}{5x} = 1$$

A fraction with a zero numerator is equal to zero.

Ex. $\frac{0}{1} = 0$, $\frac{0}{125} = 0$, $\frac{0}{8x} = 0$

A fraction with a zero denominator is called undefined.

Ex. $\frac{1}{0} = undefined$, $\frac{8x}{0} = undefined$

- 6 -

Retro-Lesson C

To Simplify or Reduce Fractions

Fractions can be simplified (or reduced) if they have a factor in common (or G.C.F., Greatest Common Factor), then the numerator and denominator or both divided by the common factor to create an equivalent fraction in reduced form.

$$reduce \; \frac{8}{12} \Rightarrow 8 \; and \; 12 \; have \; 4 \; in \; common \; as \; a \; factor \; so, \Rightarrow \frac{8 \div 4}{12 \div 4} \Rightarrow \frac{2}{3}$$

Change Mix Number to Improper Fraction

$$whole \; number \; \frac{numerator}{denominator} \longmapsto \frac{denominator \cdot whole \; number \; + \; numerator}{same \; denominator}$$

$$5\frac{2}{3} \longmapsto \frac{3 \cdot 5 + 2}{3} \longmapsto \frac{17}{3}$$

Change Improper Fraction to Mix Number

$$denominator\sqrt{numerator}^{\,quotient, remainder} \longmapsto quotient \frac{remainder}{same \; denominator}$$

$$\frac{17}{3} \longmapsto 3\sqrt{17}^{\;5 \; r2} \longmapsto 5\frac{2}{3}$$
$$\frac{15}{2}$$

- 8 -

Session Comprehension

Date: _____

1. List the retro lessons that you have difficulty in comprehending.

2. What background concepts do you need to understand in order to add, subtract, multiply and divide fractions?

3. What new concepts and methods do you anticipate learning?

4. What questions do you have regarding the retro lessons?

Addition

of

Fractions

and

Mixed Numbers

Overview of Adding Fractions with <u>like</u> denominators

Example: Add $\dfrac{1}{8} + \dfrac{3}{8}$

Add: $\dfrac{1}{8} + \dfrac{3}{8}$

Step 1. Add the numerators.

Step 2. The denominators stay the same.

$$\frac{1}{8} + \frac{3}{8} = \frac{1+3}{8} = \frac{4}{8}$$

Step 3. Simplify result, if possible.

4 is the greatest common factor of 4 and 8. $4 \times 1 = 4$
$4 \times 2 = 8$

$$\frac{4 \div 4}{8 \div 4} \Rightarrow \frac{1}{2}$$

Exercise #1. Add the following fractions.

1. $\dfrac{2}{7} + \dfrac{3}{7}$	2. $\dfrac{1}{5} + \dfrac{3}{5}$
3. $\dfrac{5}{12} + \dfrac{1}{12}$	4. $\dfrac{5}{18} + \dfrac{7}{18}$

- 14 -

Session Comprehension

Date: _____

1. List steps or definitions that you have difficulty in comprehending.

2. Explain what is most important to understand about the addition of fractions with the same denominators?

3. Explain in your own words the meaning of the term, like denominators?

4.

Illustrate solving the addition problem.	Explain each step of the process.
$\frac{4}{9} + \frac{2}{9}$	

Overview of Adding Fractions with <u>unlike</u> denominators

Example: Add $\quad \dfrac{2}{3} + \dfrac{1}{5}$

Step 1. Find the Least Common Denominator (L.C.D.).

$$3 \Rightarrow 3, 6, 9, 12, \boxed{15}$$

$$5 \Rightarrow 5, 10, \boxed{15}$$

Stop adding multiples when match is found!

The L.C.D. of 3 and 5 is 15.

Step 2. Find the Equivalent Fractions using the L.C.D.

Re-write addition vertically.

$$\dfrac{2}{3} = \dfrac{\square}{15}$$

Divide 3 into 15 to get 5.

Divide 5 into 15 to get 3.

$$\dfrac{1}{5} = \dfrac{\square}{15}$$

Multiply numerator and denominator times 3 to find equivalent numerator.

$$\dfrac{2 \cdot 5}{3 \cdot 5} = \dfrac{10}{15}$$

Multiply numerator and denominator times 5 to find equivalent numerator.

$$\dfrac{1 \cdot 3}{5 \cdot 3} = \dfrac{3}{15}$$

Step 3.
Add the numerators and simplify result, if possible.

$$\dfrac{10 + 3}{15} \Rightarrow \dfrac{13}{15}$$

$\dfrac{13}{15}$ <u>cannot</u> be simplified.

Common Denominator

A Common Denominator is the smallest matching multiple of two or more numbers.

Select the largest denominator as the common denominator, if the smaller denominator(s) divides evenly into the Largest denominator.

Find the common denominator of $\frac{2}{3}$ and $\frac{5}{6}$

> 3 divides evenly into 6, therefore the Common denominator is 6.

Find the common denominator of $\frac{2}{3}$, $\frac{5}{7}$ and $\frac{11}{21}$

> Both 3 and 7 divide evenly into 21 therefore the Common denominator is 21.

Multiply the denominators to find the common denominators.

Find the common denominator of $\frac{2}{3}$ and $\frac{5}{8}$

> Multiply 3 and 8 to get 24 as the common Denominator.

Exercise #2. Find the Least Common Denominator (L.C.D.).

1. $\frac{5}{7}, \frac{2}{5}$	2. $\frac{3}{8}, \frac{1}{3}$	3. $\frac{3}{4}, \frac{7}{12}$
4. $\frac{1}{7}, \frac{7}{21}$	5. $\frac{5}{11}, \frac{7}{33}$	6. $\frac{9}{14}, \frac{23}{42}$
7. $\frac{1}{4}, \frac{3}{5}, \frac{9}{20}$	8. $\frac{9}{14}, \frac{8}{35}, \frac{11}{70}$	9. $\frac{2}{3}, \frac{4}{17}, \frac{7}{51}$

- 17 -

Session Comprehension

Date: _____

1. List steps or definitions that you have difficulty in comprehending.

2. Explain what is most important to understand when finding like denominators for fractions with unlike denominators?

3. Explain in your own words two different methods in finding the common denominator of fractions with unlike denominators.

4.

Illustrate finding the common denominator for the following fractions.	Explain each step of the process.
$\frac{6}{7}$ and $\frac{2}{3}$ $\Longrightarrow$	
$\frac{1}{4}$ and $\frac{5}{16}$ $\Longrightarrow$	

Least Common Denominator

Least Common Denominator (L.C.D.)
The smallest matching multiply of two or more Denominators.

Prime Factorization
The expression of a number written as the multiplication of its prime factors.

<mark>Example using multiples</mark>

Find the L.C.D. of $\dfrac{3}{4}, \dfrac{1}{6}, \dfrac{5}{18}$

$4 \Rightarrow 4, 8, 12, 16, 20, 24, 28, 32, \boxed{36}$
$6 \Rightarrow 6, 12, 18, 24, 30, \boxed{36}, 42$
$18 \Rightarrow 18, \boxed{36}$

L.C.D is 36

<mark>Example using prime factorization</mark>

Find the L.C.D. of $\dfrac{3}{4}, \dfrac{1}{6}, \dfrac{5}{18}$

a. Prime Factorize each denominator.
b. Select the largest group of each prime.
c. Multiply selection.
Note: Prime set can only be selected once.

$4 \quad \Rightarrow \boxed{2 \cdot 2}$
$6 \quad \Rightarrow 2 \cdot 3$
$18 \quad \Rightarrow 2 \cdot \boxed{3 \cdot 3}$

L.C.D. is $2 \cdot 2 \cdot 3 \cdot 3 = 36$

Exercise #3. Find the Least Common Denominator (L.C.D.).

1. $\dfrac{5}{10}, \dfrac{3}{8}$	2. $\dfrac{3}{8}, \dfrac{1}{14}$	3. $\dfrac{3}{5}, \dfrac{7}{12}$
4. $\dfrac{1}{6}, \dfrac{7}{8}$	5. $\dfrac{5}{12}, \dfrac{11}{18}$	6. $\dfrac{2}{3}, \dfrac{9}{16}$
7. $\dfrac{1}{3}, \dfrac{2}{5}, \dfrac{9}{10}$	8. $\dfrac{3}{10}, \dfrac{8}{15}, \dfrac{11}{25}$	9. $\dfrac{3}{4}, \dfrac{4}{5}, \dfrac{7}{8}$

Session Comprehension

Date: _____

1. List steps or definitions that you have difficulty in comprehending.

2. Explain what is most important to understand when finding <u>least common denominators</u> for fractions with unlike denominators.

3. Explain in your own words two different methods in finding the <u>least common denominator</u> for fractions with unlike denominators.

4.

Illustrate two methods in finding the <u>least common denominator</u> for the following fractions.	Explain each step of the process.
$\frac{7}{18}$ and $\frac{3}{4}$	
$\frac{7}{18}$ and $\frac{3}{4}$	

How do we find the numerators of equivalent fractions?

Step 1. Divide the denominator into the equivalent denominator to generate a factor.

$$denominator \sqrt{\overset{\textstyle factor}{equivalent\ denominator}}$$

Step 2. Multiply the factor times the numerator to generate the equivalent numerator.

$$\frac{numerator \cdot factor}{denominator \cdot factor} = \frac{equivalent\ numerator}{equivalent\ denominator}$$

Example 1.

Find the equivalent numerator: $\dfrac{3}{5} = \dfrac{\blacksquare}{10}$

1. Divide **5** into **10** and get a factor of **2**.

2. Multiply the factor times the numerator.

For consistency show the numerator and denominator being multiplied by the *factor*.

$$\frac{3 \cdot 2}{5 \cdot 2} = \frac{6}{10}$$

Example 2.

Find the equivalent numerator: $4 = \dfrac{\square}{5}$

1. Turn the whole number into an improper fraction by placing it over 1.

$$\frac{4}{1} = \frac{\square}{5}$$

2. Divide 1 into 5 to get a factor of 5.
3. Multiply the numerator and denominator times 5

$$\frac{4 \cdot 5}{1 \cdot 5} = \frac{20}{5}$$

Exercise #4. Find the equivalent numerators.

1. $\dfrac{1}{2} = \dfrac{[\ \ ]}{8}$	2. $\dfrac{5}{6} = \dfrac{[\ \ ]}{24}$
3. $\dfrac{4}{17} = \dfrac{[\ \ ]}{51}$	4. $\dfrac{7}{12} = \dfrac{[\ \ ]}{36}$
5. $3 = \dfrac{[\ \ ]}{6}$	6. $5 = \dfrac{[\ \ ]}{7}$

- 22 -

Session Comprehension

Date: _____

1. List steps or definitions that you have difficulty in comprehending.

2. Explain what is most important to understand when creating an equivalent fraction.

3. Explain in your own words the meaning of equivalent fractions.

4.

Illustrate creating equivalent numerators for the following fraction and whole number.	Explain each step of the process.
$$\frac{5}{9} = \frac{[\ \ ]}{36}$$	
$$4 = \frac{[\ \ ]}{6}$$	

Summary of Adding Fractions

Example: Add $\dfrac{3}{7} + \dfrac{5}{21}$

Step 1. <u>In order to add fractions the denominators must be the same</u>, so find the Least Common Denominator (L.C.D.).

7 divides evenly into 21, so the L.C.D. is 21.

Step 2. Find the <u>Equivalent Fractions</u> using the L.C.D.
Re-write addition vertically.

$$\frac{3}{7} = \frac{\square}{21}$$

Divide 7 into 21 to get 3.

Divide 21 into 21 to get 1.

$$\frac{5}{21} = \frac{\square}{21}$$

$$\frac{3 \cdot 3}{7 \cdot 3} = \frac{9}{21}$$

Multiply numerator and denominator times 3 to find equivalent numerator.

Multiply numerator and denominator times 1 to find equivalent numerator.

$$\frac{5 \cdot 1}{27 \cdot 1} = \frac{5}{21}$$

$$\frac{14 \div 7}{27 \div 7} \Rightarrow \frac{2}{3}$$

Step 3.

Add the numerators and simplify result, if possible.

$$\frac{9 + 5}{21} \Rightarrow \frac{14}{21} \Rightarrow \frac{2}{3}$$

Session Comprehension

Date: _____

1. List steps or definitions that you have difficulty in comprehending.

2. Explain what is most important to understand when adding fraction with unlike denominators.

3. Explain in your own words when and why it is necessary to create equivalent fractions when adding fractions.

4.

Illustrate adding the following fractions.	Explain each step of the process.
$\frac{1}{8} + \frac{2}{9}$	

Exercise #5. Add the following fractions.

1. $\dfrac{1}{4} + \dfrac{5}{12}$	2. $\dfrac{3}{7} + \dfrac{2}{5}$
3. $\dfrac{3}{8} + \dfrac{1}{3}$	4. $\dfrac{1}{7} + \dfrac{7}{21}$
5. $\dfrac{5}{11} + \dfrac{7}{33}$	6. $\dfrac{2}{3} + \dfrac{3}{16}$
7. $\dfrac{2}{3} + \dfrac{4}{17} + \dfrac{2}{51}$	8. $\dfrac{1}{4} + \dfrac{2}{5} + \dfrac{3}{20}$

- 26 -

Adding Mixed Numbers

Example: Add $\quad 5\dfrac{2}{3} + 6\dfrac{1}{4}$

How do we add mixed numbers?

Step 1. Find the L.C.D. of the fractions.

The L.C.D. of 3 and 4 is 12.

Step 2. Create Equivalent Fractions using the L.C.D.
Re-write addition vertically.

$$5\,\dfrac{2 \cdot 4}{3 \cdot 4} = \dfrac{8}{12}$$

Divide 3 into 12 to get 4.

$$6\,\dfrac{1 \cdot 3}{4 \cdot 3} = \dfrac{3}{12}$$

Divide 4 into 12 to get 3.

Step 3. Add the whole numbers, then add the numerators and simplify result, if possible.

$$5\,\dfrac{8}{12}$$

$$6\,\dfrac{3}{12}$$

$$11\,\dfrac{11}{12}$$

$5 + 6 = 11$

$\dfrac{8 + 3}{12} = \dfrac{11}{12}$

- 27 -

Session Comprehension

Date: _____

1. List steps or definitions that you have difficulty in comprehending.

2. Explain what is most important to understand when adding mixed numbers.

3. Describe in your own words what makes a mixed number a mixed number.

4.

Illustrate adding the following mixed numbers.	Explain each step of the process.
$3\frac{1}{3} + 7\frac{1}{5}$	

Exercise #6. Add the following mixed numbers.

1. $2\frac{1}{2} + 1\frac{1}{3}$	2. $4\frac{2}{3} + 1\frac{2}{15}$
3. $2\frac{3}{8} + 9\frac{1}{2}$	4. $5\frac{2}{10} + 10\frac{1}{3}$
5. $5\frac{2}{9} + 7\frac{3}{4}$	6. $7\frac{3}{8} + 9\frac{1}{6} + 3\frac{1}{12}$

- 29 -

Adding a Whole Number to a Mixed Number

Example: Add $\quad 4\dfrac{3}{8} + 5$

How do we add a whole number to a mixed number?

Step 1. Add the whole numbers and keep the fraction.

Re-write addition vertically.

$$
\begin{array}{r}
4\dfrac{3}{8} \\
5 \\
\hline
9\dfrac{3}{8}
\end{array}
$$

Add the whole numbers 4 and 5.

Bring down the fraction.

Result (answer) with an Improper fraction

Example: Add $\quad 4\dfrac{3}{4} + 5\dfrac{2}{3} = 9\dfrac{17}{12}$

Step 1. Change the improper fraction to a mixed number and add the whole numbers.

$$9\dfrac{17}{12} \Rightarrow 9 + \dfrac{17}{12} \Rightarrow 9 + 1\dfrac{5}{12} \Rightarrow 10\dfrac{5}{12}$$

$\dfrac{17}{12}$ *becomes* $1\dfrac{5}{12}$ *as a mixed number*

Session Comprehension

Date: _____

1. List steps or definitions that you have difficulty in comprehending.

2. Explain what is most important to understand when adding a mixed number to a whole number.

3. Describe in your own words the steps needed to change a mixed number with an improper fraction to a mixed number with a proper fraction.

4.

Illustrate adding the following mixed numbers.	Explain each step of the process.
$$6\frac{5}{6} + 17\frac{2}{9}$$	

Exercise #7.   Add the following mixed numbers.

1. $2\frac{1}{3} + 5$	2. $3 + 9\frac{4}{5}$
3. $6\frac{5}{8} + 2\frac{3}{4}$	4. $12\frac{4}{7} + 8\frac{8}{9}$
5. $14\frac{5}{11} + 10\frac{15}{22}$	6. $22\frac{7}{8} + 16\frac{5}{6}$

Copyright© 2010 by Wanda Y. Brooks
Igeddit Creative Concepts, LLC., All rights reserved.

Subtraction

of

Fractions

and

Mixed Numbers

Subtraction of Fractions

Example: Subtract $\dfrac{5}{6} - \dfrac{7}{12}$

How do we subtract fractions?

Step 1. Find the L.C.D. of the fractions.

The L.C.D. of 6 and 12 is 12.

Step 2. Find the Equivalent Fractions using the L.C.D.
Re-write addition vertically.

$$\frac{5 \cdot 2}{6 \cdot 2} = \frac{10}{12}$$

Divide 6 into 12 to get 2.

Divide 12 into 12 to get 1.

$$\frac{7 \cdot 1}{12 \cdot 1} = \frac{7}{12}$$

Step 3. <u>Subtract</u> the numerators and simplify result, if possible.

$$\frac{10}{12}$$
$$\frac{7}{12}$$

$$\frac{10 - 7}{12} \Rightarrow \frac{3}{12} \Rightarrow \frac{1}{4}$$

Simplify result by a factor of 3

$$\frac{3 \div 3}{12 \div 3} \Rightarrow \frac{1}{4}$$

Session Comprehension

Date: _____

1. List steps or definitions that you have difficulty in comprehending.

2. Explain what is most important to understand when subtracting fractions.

3. Describe in your own words the one step that is different when adding fractions in contrast to subtracting fractions.

4.

Illustrate subtracting the following fractions.	Explain each step of the process.
$$\frac{5}{8} - \frac{7}{24}$$	

- 37 -

Exercise #8. Subtract the following fractions.

1. $\dfrac{5}{12} - \dfrac{2}{12}$	2. $\dfrac{9}{10} - \dfrac{3}{10}$
3. $\dfrac{1}{3} - \dfrac{1}{5}$	4. $\dfrac{5}{6} - \dfrac{9}{24}$
5. $\dfrac{5}{8} - \dfrac{1}{6}$	6. $\dfrac{11}{12} - \dfrac{3}{4}$
7. $\dfrac{7}{12} - \dfrac{12}{30}$	8. $\dfrac{5}{6} - \dfrac{2}{5}$

Subtracting Mixed Numbers

Example: Subtract $9\dfrac{7}{15} - 7\dfrac{2}{25}$

How do we subtract mixed numbers?

Step 1. Find the L.C.D. of the fractions.

The L.C.D. of **15** and **25** is **75**.

Step 2. Find the Equivalent Fractions using the L.C.D.

Re-write addition vertically.

$$9\dfrac{7\cdot 5}{15\cdot 5} = \dfrac{35}{75}$$

Divide 15 into 75 to get 5.

Divide 25 into 75 to get 3.

$$7\dfrac{2\cdot 3}{25\cdot 3} = \dfrac{6}{75}$$

Step 3. Subtract the whole numbers, then subtract the numerators and simplify result, if possible.

$$9\;\dfrac{35}{75}$$
$$6\;\dfrac{6}{75}$$
$$\overline{}$$
$$3\;\dfrac{29}{75}$$

- 39 -

Session Comprehension

Date: _____

1. List steps or definitions that you have difficulty in comprehending.

2. Explain what is most important to understand when subtracting mixed numbers.

3. Describe in your own words the similarities in steps when adding mixed numbers and subtracting mixed numbers.

4.

Illustrate subtracting the following mixed numbers.	Explain each step of the process.
$$5\frac{7}{12} - 2\frac{4}{9}$$	

Exercise #9. Subtract the following mixed numbers.

1. $19\frac{7}{8} - 7\frac{5}{8}$

2. $23\frac{16}{17} - 11\frac{12}{17}$

3. $2\frac{1}{2} - 1\frac{1}{3}$

4. $4\frac{2}{3} - 1\frac{1}{5}$

5. $58\frac{8}{35} - 12\frac{3}{14}$

6. $15\frac{3}{5} - 8\frac{4}{9}$

Subtracting a Whole Number from a Mixed Number

Example: Subtract $7\dfrac{3}{5} - 4$

How do we subtract a mixed number from a whole number?

Step 1. Subtract the whole numbers.

Step 2. Bring down the fraction.

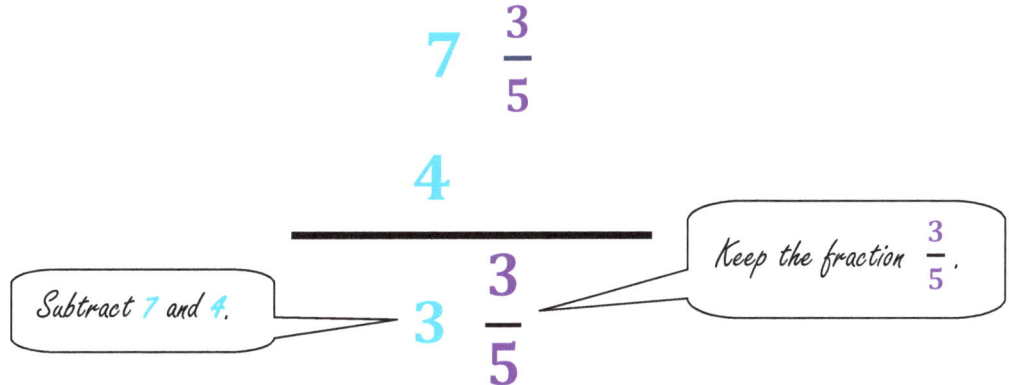

$$7 \quad \frac{3}{5}$$
$$4$$
$$\overline{}$$
$$3 \quad \frac{3}{5}$$

Subtract 7 and 4.

Keep the fraction $\dfrac{3}{5}$.

Exercise #10. Subtract the following.

1. $6\dfrac{1}{5} - 3$	2. $15\dfrac{7}{8} - 4$
3. $18\dfrac{7}{18} - 7$	4. $102\dfrac{11}{20} - 75$

- 42 -

Session Comprehension

Date: _____

1. List steps or definitions that you have difficulty in comprehending.

2. Explain what is most important to understand when subtracting a whole number from a mixed number.

3. Explain in your own words the process of subtracting a whole number from a mixed number.

4.

Illustrate subtracting the following mixed number and whole number.	Explain each step of the process.
$6\frac{3}{8} - 4$	

Borrowing: Subtracting a Mixed Number from a Whole Number

Example: Subtract $5 - 2\dfrac{4}{11}$

How do we turn a whole number into a mixed number?

Step 1. Borrow 1 from the whole number.

5 becomes $4 + 1$

Step 2. Change the 1 into a fraction with <u>matching numerator and denominator</u> that is equivalent to the denominator of the mixed number.

$4 + 1$ becomes $4 + \dfrac{11}{11}$ which is the same as $4\dfrac{11}{11}$

How do we subtract a mixed number from a whole number?

Example: Subtract $5 - 2\dfrac{4}{11}$

Borrow 1 from 5, turn 1 into an equivalent fraction, then subtract whole numbers and numerators.

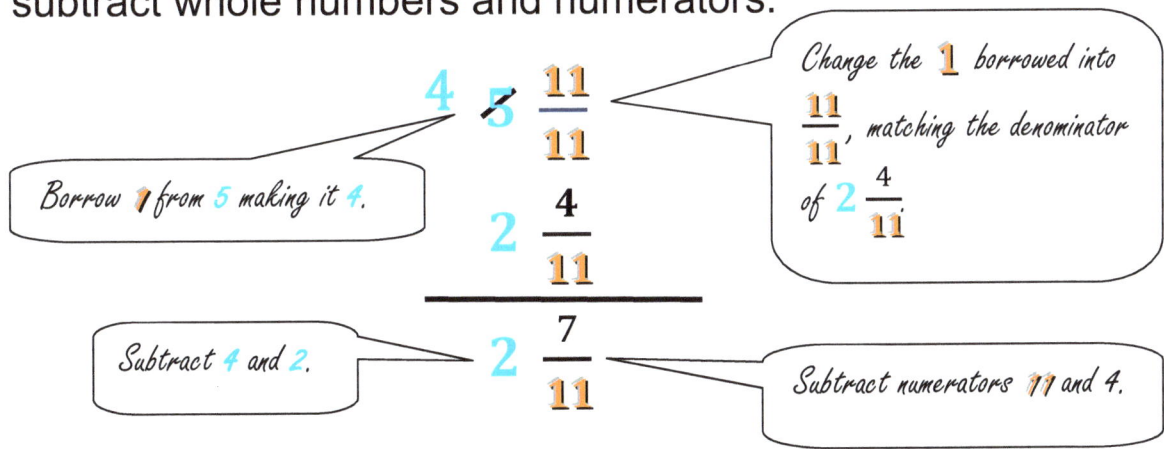

Session Comprehension

Date: _____

1. List steps or definitions that you have difficulty in comprehending.

2. Explain what is most important to understand when subtracting a mixed number from a whole number.

3. Explain in your own words why it is necessary to borrow one from the whole number when subtracting a mixed number from a whole number.

4.

Illustrate subtracting the following whole number and mixed number. $$8 - 4\frac{3}{5}$$	Explain each step of the process.

Exercise #11. Subtract the following mixed numbers.

1. $10 - 6\frac{1}{3}$	2. $3 - 2\frac{5}{51}$
3. $12 - 10\frac{7}{9}$	4. $63 - 52\frac{19}{24}$
5. $6 - 3\frac{1}{5}$	6. $15 - 4\frac{7}{8}$
7. $18 - 7\frac{7}{18}$	8. $102 - 75\frac{11}{20}$

Borrowing: Subtracting Mixed Numbers

Example: Subtract $5\dfrac{1}{4} - 3\dfrac{2}{3}$

How do we borrow in order to subtract mixed numbers?

Step 1. Find the L.C.D. of the fractions.

The L.C.D. of **4** and **3** is **12**.

Step 2. Find the Equivalent Fractions using the L.C.D.

$$5\dfrac{1 \cdot 3}{4 \cdot 3} = \dfrac{3}{12}$$

a. Divide 4 into 12 to get 3.
b. The numerator of the top fraction is smaller than the numerator of the bottom fraction.

Divide 3 into 12 to get 4.

$$3\dfrac{2 \cdot 4}{3 \cdot 4} = \dfrac{8}{12}$$

Step 3. <mark>Because the first numerator is smaller than the second numerator</mark>, one must be borrowed from the first whole number and added to its fraction.

Note: How do you borrow one from a whole number to increase the fraction?

Example: $7\dfrac{3}{5}$

a. A mixed number is the sum of a whole number and a fraction.

$7\dfrac{3}{5}$ becomes $7 + \dfrac{3}{5}$

- 47 -

b. To increase fraction size, borrow <u>one</u> from the whole number.

c. Change the <u>one</u> borrowed from the whole number into a fraction by making the numerator and denominator the <u>same as</u> the original fraction's denominator.

d. Add together the whole number and fractions.

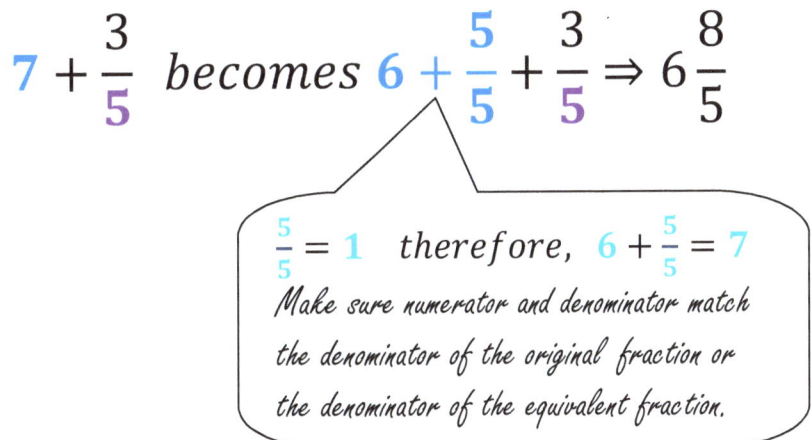

$$7 + \frac{3}{5} \quad becomes \quad 6 + \frac{5}{5} + \frac{3}{5} \Rightarrow 6\frac{8}{5}$$

$\frac{5}{5} = 1 \quad therefore, \quad 6 + \frac{5}{5} = 7$

Make sure numerator and denominator match the denominator of the original fraction or the denominator of the equivalent fraction.

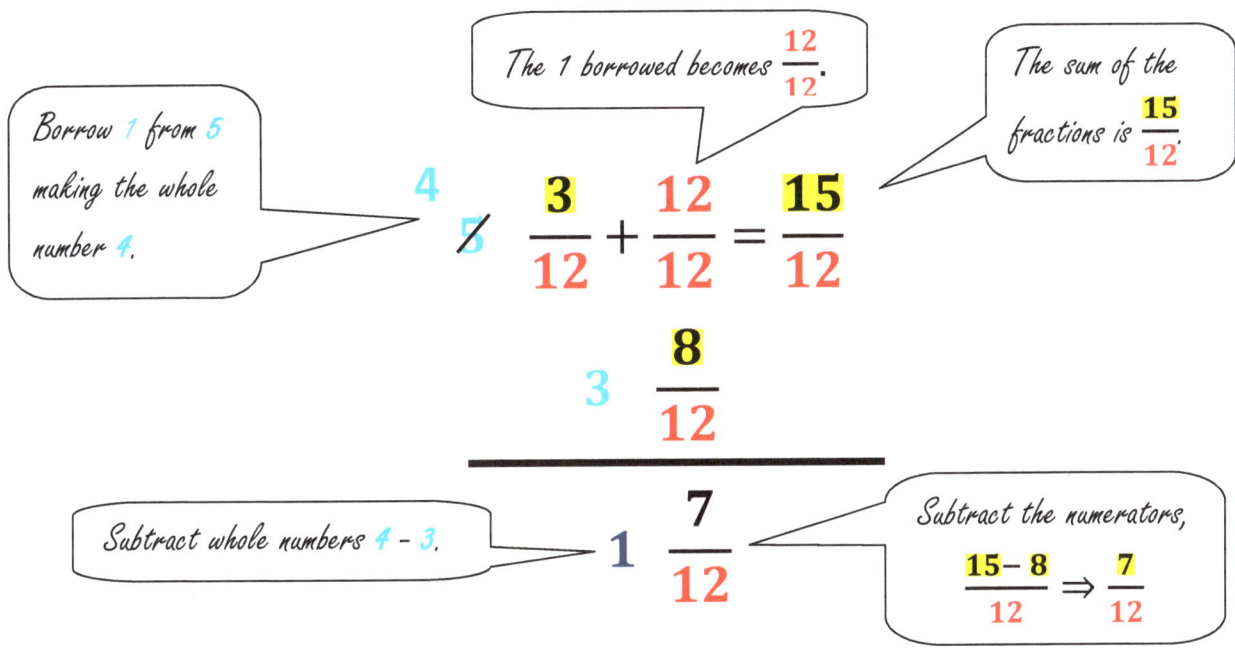

The 1 borrowed becomes $\frac{12}{12}$.

The sum of the fractions is $\frac{15}{12}$.

Borrow 1 from 5 making the whole number 4.

$$4 \quad \not{5} \quad \frac{3}{12} + \frac{12}{12} = \frac{15}{12}$$

$$3 \quad \frac{8}{12}$$

Subtract whole numbers 4 − 3.

$$1 \quad \frac{7}{12}$$

Subtract the numerators, $\frac{15-8}{12} \Rightarrow \frac{7}{12}$

Session Comprehension

Date: _____

1. List steps or definitions that you have difficulty in comprehending.

2. Explain what is most important to understand when subtracting mixed numbers.

3. Explain in your own words when it is necessary to borrow one from the whole number of a mixed number when subtracting mixed numbers.

4.

Illustrate subtracting the following whole number and mixed number.	Explain each step of the process.
$16\frac{1}{5} - 9\frac{5}{9}$	

Exercise #12. Subtract the following mixed numbers.

1. $7\frac{5}{8} - 5\frac{3}{4}$	2. $13\frac{17}{34} - 10\frac{11}{17}$
3. $12\frac{1}{4} - 11\frac{1}{3}$	4. $8\frac{3}{5} - 7\frac{3}{4}$
5. $5\frac{1}{6} - 2\frac{11}{12}$	6. $16\frac{5}{6} - 9\frac{9}{10}$
7. $21\frac{2}{3} - 14\frac{7}{8}$	8. $48\frac{2}{7} - 22\frac{13}{35}$

- 50 -

Exercise #1. Add the following fractions.

1. $\dfrac{2}{7} + \dfrac{3}{7}$

$$\frac{2}{7} + \frac{3}{7} = \frac{2+3}{7} = \frac{5}{7}$$

2. $\dfrac{1}{5} + \dfrac{3}{5}$

$$\frac{1}{5} + \frac{3}{5} = \frac{1+3}{5} = \frac{4}{5}$$

3. $\dfrac{5}{12} + \dfrac{1}{12}$

$$\frac{5}{12} + \frac{1}{12} = \frac{5+1}{12} = \frac{6}{12}$$

$$\frac{6 \div 6}{12 \div 6} = \frac{1}{2}$$

4. $\dfrac{5}{18} + \dfrac{7}{18}$

$$\frac{5}{18} + \frac{7}{18} = \frac{5+7}{18} = \frac{12}{18}$$

$$\frac{12 \div 6}{18 \div 6} = \frac{2}{3}$$

Exercise #2. Find the Least Common Denominator (L.C.D.).

1. $\dfrac{5}{7}, \dfrac{2}{5}$

$$7 \times 5 = 35$$
$$l.c.d = 35$$

2. $\dfrac{3}{8}, \dfrac{1}{3}$

$$8 \times 3 = 24$$
$$l.c.d. = 24$$

3. $\dfrac{3}{4}, \dfrac{7}{12}$

$$12 \div 4 = 3$$
$$l.c.d. = 12$$

4. $\dfrac{1}{7}, \dfrac{7}{21}$

$$21 \div 7 = 3$$
$$l.c.d. = 21$$

5. $\dfrac{5}{11}, \dfrac{7}{33}$

$$33 \div 11 = 3$$
$$l.c.d. = 33$$

6. $\dfrac{9}{14}, \dfrac{23}{42}$

$$42 \div 14 = 3$$
$$l.c.d. = 42$$

7. $\dfrac{1}{4}, \dfrac{3}{5}, \dfrac{9}{20}$

$$20 \div 4 = 5$$
$$20 \div 5 = 4$$
$$l.c.d. = 20$$

8. $\dfrac{9}{14}, \dfrac{8}{35}, \dfrac{11}{70}$

$$70 \div 14 = 5$$
$$70 \div 35 = 2$$
$$l.c.d. = 70$$

9. $\dfrac{2}{3}, \dfrac{4}{17}, \dfrac{7}{51}$

$$51 \div 3 = 17$$
$$51 \div 17 = 3$$
$$l.c.d. = 51$$

- 54 -

Exercise #3. Find the Least Common Denominator (L.C.D.).

1. $\dfrac{5}{10}, \dfrac{3}{8}$ $10 - 10, 20, 30, 40$ $8 - 8, 16, 24, 32, 40$ $l.c.d = 40$	2. $\dfrac{3}{8}, \dfrac{1}{14}$ $8 - 8, 16, 24, 32, 40, 48, 56$ $14 - 14, 28, 42, 56$ $l.c.d = 56$	3. $\dfrac{3}{5}, \dfrac{7}{12}$ $5 - 5$ $12 - 2 \cdot 2 \cdot 3$ $l.c.d = 2 \cdot 2 \cdot 3 \cdot 5 = 60$
4. $\dfrac{1}{6}, \dfrac{7}{8}$ $6 - 6, 12, 18, 24$ $8 - 8, 16, 24$ $l.c.d = 24$	5. $\dfrac{5}{12}, \dfrac{11}{18}$ $12 - 12, 24, 36$ $18 - 18, 36$ $l.c.d = 36$	6. $\dfrac{2}{3}, \dfrac{9}{16}$ $3 - 3$ $12 - 2 \cdot 2 \cdot 2 \cdot 2$ $l.c.d = 2 \cdot 2 \cdot 2 \cdot 2 \cdot 3 = 48$
7. $\dfrac{1}{3}, \dfrac{2}{5}, \dfrac{9}{10}$ $3 - 3, 6, 9, 12, 15, 18, 21, 24$ $\quad 27, 30$ $5 - 5, 10, 15, 20, 25, 30$ $10 - 10, 20, 30$ $l.c.d = 30$	8. $\dfrac{3}{10}, \dfrac{8}{15}, \dfrac{11}{25}$ $10 - 2 \cdot 5$ $15 - 3 \cdot 5$ $25 - 5 \cdot 5$ $l.c.d. = 2 \cdot 3 \cdot 5 \cdot 5 = 150$	9. $\dfrac{3}{4}, \dfrac{4}{5}, \dfrac{7}{8}$ $4 - 2 \cdot 2$ $5 - 5$ $8 - 2 \cdot 2 \cdot 2$ $l.c.d. = 2 \cdot 2 \cdot 2 \cdot 5 = 40$

Exercise #4. Find the equivalent numerators.

$\dfrac{1}{2} = \dfrac{[4]}{8}$	$\dfrac{5}{6} = \dfrac{[20]}{24}$
$\dfrac{4}{17} = \dfrac{[12]}{51}$	$\dfrac{7}{12} = \dfrac{[21]}{36}$
$3 = \dfrac{[18]}{6}$	$5 = \dfrac{[35]}{7}$

- 55 -

Exercise #5. Add the following fractions.

1. $\dfrac{1}{4} + \dfrac{5}{12}$

$$\dfrac{1}{4} = \dfrac{3}{12}$$

$$\dfrac{5}{12} = \dfrac{5}{12}$$

$$\dfrac{8}{12} \rightarrow \dfrac{8 \div 4}{12 \div 4} \rightarrow \dfrac{2}{3}$$

2. $\dfrac{3}{7} + \dfrac{2}{5}$

$$\dfrac{3}{7} = \dfrac{15}{35}$$

$$\dfrac{2}{5} = \dfrac{14}{35}$$

$$\dfrac{29}{35}$$

3. $\dfrac{3}{8} + \dfrac{1}{3}$

$$\dfrac{3}{8} = \dfrac{9}{24}$$

$$\dfrac{1}{3} = \dfrac{8}{24}$$

$$\dfrac{17}{24}$$

4. $\dfrac{1}{7} + \dfrac{7}{21}$

$$\dfrac{1}{7} = \dfrac{3}{21}$$

$$\dfrac{7}{21} = \dfrac{7}{21}$$

$$\dfrac{10}{21}$$

5. $\dfrac{5}{11} + \dfrac{7}{33}$

$$\dfrac{5}{11} = \dfrac{15}{33}$$

$$\dfrac{7}{33} = \dfrac{7}{33}$$

$$\dfrac{22}{33} \rightarrow \dfrac{22 \div 11}{33 \div 11} \rightarrow \dfrac{2}{3}$$

6. $\dfrac{2}{3} + \dfrac{3}{16}$

$$\dfrac{2}{3} = \dfrac{32}{48}$$

$$\dfrac{3}{16} = \dfrac{9}{48}$$

$$\dfrac{41}{48}$$

- 56 -

Exercise #5. Add the following fractions. Continued...

7. $\dfrac{2}{3} + \dfrac{4}{17} + \dfrac{2}{51}$

$$\dfrac{2}{3} = \dfrac{34}{51}$$

$$\dfrac{4}{17} = \dfrac{12}{51}$$

$$\dfrac{2}{51} = \dfrac{2}{51}$$

$$\dfrac{48}{51} \rightarrow \dfrac{48 \div 3}{51 \div 3} \rightarrow \dfrac{16}{17}$$

8. $\dfrac{1}{4} + \dfrac{2}{5} + \dfrac{3}{20}$

$$\dfrac{1}{4} = \dfrac{5}{20}$$

$$\dfrac{2}{5} = \dfrac{8}{20}$$

$$\dfrac{3}{20} = \dfrac{3}{20}$$

$$\dfrac{16}{20} \rightarrow \dfrac{16 \div 4}{20 \div 4} \rightarrow \dfrac{4}{5}$$

Exercise #6. Add the following mixed numbers.

1. $2\frac{1}{2} + 1\frac{1}{3}$

$$2\frac{1}{2} = \frac{3}{6}$$

$$1\frac{1}{3} = \frac{2}{6}$$

$$\rule{3cm}{0.4pt}$$

$$3\frac{5}{6}$$

2. $4\frac{2}{3} + 1\frac{2}{15}$

$$4\frac{2}{3} = \frac{10}{15}$$

$$1\frac{2}{15} = \frac{2}{15}$$

$$\rule{3cm}{0.4pt}$$

$$5\frac{12 \div 3}{15 \div 3} \rightarrow 5\frac{4}{5}$$

3. $2\frac{3}{8} + 9\frac{1}{2}$

$$2\frac{3}{8} = \frac{3}{8}$$

$$9\frac{1}{2} = \frac{4}{8}$$

$$\rule{3cm}{0.4pt}$$

$$11\frac{7}{8}$$

4. $5\frac{2}{10} + 10\frac{1}{3}$

$$5\frac{2}{10} = \frac{6}{30}$$

$$10\frac{1}{3} = \frac{10}{30}$$

$$\rule{3cm}{0.4pt}$$

$$15\frac{16}{30} \rightarrow 15\frac{8}{15}$$

5. $5\frac{2}{9} + 7\frac{3}{4}$

$$5\frac{2}{9} = \frac{8}{36}$$

$$7\frac{3}{4} = \frac{27}{36}$$

$$\rule{3cm}{0.4pt}$$

$$12\frac{35}{36}$$

6. $7\frac{3}{8} + 9\frac{1}{6} + 3\frac{1}{12}$

$$7\frac{3}{8} = \frac{9}{24}$$

$$9\frac{1}{6} = \frac{4}{24}$$

$$3\frac{1}{12} = \frac{2}{24}$$

$$\rule{3cm}{0.4pt}$$

$$19\frac{15}{24} \rightarrow 19\frac{5}{8}$$

- 58 -

Exercise #7. Add the following mixed numbers.

1. $2\frac{1}{3} + 5$

$$2\frac{1}{3}$$

$$5$$

$$\rule{3cm}{0.4pt}$$

$$7\frac{1}{3}$$

2. $3 + 9\frac{4}{5}$

$$3$$

$$9\frac{4}{5}$$

$$\rule{3cm}{0.4pt}$$

$$12\frac{4}{5}$$

3. $6\frac{5}{8} + 2\frac{3}{4}$

$$6\frac{5}{8} = \frac{5}{8}$$

$$2\frac{3}{4} = \frac{6}{8}$$

$$\rule{3cm}{0.4pt}$$

$$8\frac{11}{8} \rightarrow 8 + 1\frac{3}{8} \rightarrow 9\frac{3}{8}$$

4. $12\frac{4}{7} + 8\frac{8}{9}$

$$12\frac{4}{7} = \frac{36}{63}$$

$$8\frac{8}{9} = \frac{56}{63}$$

$$\rule{3cm}{0.4pt}$$

$$20\frac{92}{63} \rightarrow 20 + 1\frac{29}{63} \rightarrow 21\frac{29}{63}$$

5. $14\frac{5}{11} + 10\frac{15}{22}$

$$14\frac{5}{11} = \frac{10}{22}$$

$$10\frac{15}{22} = \frac{15}{22}$$

$$\rule{3cm}{0.4pt}$$

$$24\frac{25}{22} \rightarrow 24 + 1\frac{3}{22} \rightarrow 25\frac{3}{22}$$

6. $22\frac{7}{8} + 16\frac{5}{6}$

$$22\frac{7}{8} = \frac{21}{24}$$

$$16\frac{5}{6} = \frac{20}{24}$$

$$\rule{3cm}{0.4pt}$$

$$38\frac{41}{24} \rightarrow 38 + 1\frac{17}{24} \rightarrow 39\frac{17}{24}$$

- 59 -

Exercise #8. Subtract the following fractions.

1. $\dfrac{5}{12} - \dfrac{2}{12}$

$$\dfrac{5}{12}$$
$$- \quad \dfrac{2}{12}$$

$$\dfrac{5-2}{12} \rightarrow \dfrac{3}{12} \rightarrow \dfrac{3 \div 3}{12 \div 3} \rightarrow \dfrac{1}{4}$$

2. $\dfrac{9}{10} - \dfrac{3}{10}$

$$\dfrac{9}{10}$$
$$- \quad \dfrac{3}{10}$$

$$\dfrac{6}{10} \rightarrow \dfrac{3}{5}$$

3. $\dfrac{1}{3} - \dfrac{1}{5}$

$$\dfrac{1}{3} = \dfrac{5}{15}$$
$$- \quad \dfrac{1}{5} = \dfrac{3}{15}$$

$$\dfrac{2}{15}$$

4. $\dfrac{5}{6} - \dfrac{9}{24}$

$$\dfrac{5}{6} = \dfrac{20}{24}$$
$$- \quad \dfrac{9}{24} = \dfrac{9}{24}$$

$$\dfrac{11}{24}$$

5. $\dfrac{5}{8} - \dfrac{1}{6}$

$$\dfrac{5}{8} = \dfrac{15}{24}$$
$$- \quad \dfrac{1}{6} = \dfrac{4}{24}$$

$$\dfrac{11}{24}$$

6. $\dfrac{11}{12} - \dfrac{3}{4}$

$$\dfrac{11}{12} = \dfrac{11}{12}$$
$$- \quad \dfrac{3}{4} = \dfrac{9}{12}$$

$$\dfrac{2}{12} \rightarrow \dfrac{1}{6}$$

7. $\dfrac{7}{12} - \dfrac{12}{30}$

$$\dfrac{7}{12} = \dfrac{35}{60}$$
$$- \quad \dfrac{12}{30} = \dfrac{24}{60}$$

$$\dfrac{11}{60}$$

8. $\dfrac{5}{6} - \dfrac{2}{5}$

$$\dfrac{5}{6} = \dfrac{25}{30}$$
$$- \quad \dfrac{2}{5} = \dfrac{12}{30}$$

$$\dfrac{13}{30}$$

- 60 -

Exercise #9. Subtract the following mixed numbers.

1. $19\frac{7}{8} - 7\frac{5}{8}$

$$19\frac{7}{8}$$

$$7\frac{5}{8}$$

$$12\frac{2}{8} \rightarrow 12\frac{1}{4}$$

2. $23\frac{16}{17} - 11\frac{12}{17}$

$$23\frac{16}{17}$$

$$11\frac{12}{17}$$

$$12\frac{4}{17}$$

3. $2\frac{1}{2} - 1\frac{1}{3}$

$$2\frac{1}{2} = \frac{3}{6}$$

$$1\frac{1}{3} = \frac{2}{6}$$

$$1\frac{1}{6}$$

4. $4\frac{2}{3} - 1\frac{1}{5}$

$$4\frac{2}{3} = \frac{10}{15}$$

$$1\frac{1}{5} = \frac{3}{15}$$

$$3\frac{7}{15}$$

5. $58\frac{8}{35} - 12\frac{3}{14}$

$$58\frac{8}{35} = \frac{16}{70}$$

$$12\frac{3}{14} = \frac{15}{70}$$

$$46\frac{1}{70}$$

6. $15\frac{3}{5} - 8\frac{4}{9}$

$$15\frac{3}{5} = \frac{27}{45}$$

$$8\frac{4}{9} = \frac{20}{45}$$

$$7\frac{7}{45}$$

Exercise #10. Subtract the following.

1. $6\frac{1}{5} - 3$

$$6\frac{1}{5}$$
$$3$$
$$\overline{}\quad\overline{\phantom{3\frac{1}{5}}}$$
$$3\frac{1}{5}$$

2. $15\frac{7}{8} - 4$

$$15\frac{7}{8}$$
$$4$$
$$\overline{}\quad\overline{\phantom{11\frac{7}{8}}}$$
$$11\frac{7}{8}$$

3. $18\frac{7}{18} - 7$

$$18\frac{7}{18}$$
$$7$$
$$\overline{}\quad\overline{\phantom{11\frac{7}{18}}}$$
$$11\frac{7}{18}$$

4. $102\frac{11}{20} - 75$

$$102\frac{11}{20}$$
$$75$$
$$\overline{}\quad\overline{\phantom{77\frac{11}{20}}}$$
$$77\frac{11}{20}$$

Exercise #11. Subtract the following mixed numbers.

1. $10 - 6\frac{1}{3}$

$$10 \to 9\frac{3}{3}$$

$$-\quad 6\frac{1}{3}$$

$$\rule{2cm}{0.4pt}$$

$$3\frac{2}{3}$$

2. $3 - 2\frac{5}{51}$

$$3 \to 2\frac{51}{51}$$

$$-\quad 2\frac{5}{51}$$

$$\rule{2cm}{0.4pt}$$

$$\frac{46}{51}$$

3. $12 - 10\frac{7}{9}$

$$12 \to 11\frac{9}{9}$$

$$-\quad 10\frac{7}{9}$$

$$\rule{2cm}{0.4pt}$$

$$1\frac{2}{9}$$

4. $63 - 52\frac{19}{24}$

$$63 \to 62\frac{24}{24}$$

$$-\quad 52\frac{19}{24}$$

$$\rule{2cm}{0.4pt}$$

$$10\frac{5}{24}$$

5. $6 - 3\frac{1}{5}$

$$6 \to 5\frac{5}{5}$$

$$-\quad 3\frac{1}{5}$$

$$\rule{2cm}{0.4pt}$$

$$2\frac{4}{5}$$

6. $15 - 4\frac{7}{8}$

$$10 \to 9\frac{3}{3}$$

$$-\quad 6\frac{1}{3}$$

$$\rule{2cm}{0.4pt}$$

$$3\frac{2}{3}$$

7. $18 - 7\frac{7}{18}$

$$18 \to 17\frac{18}{18}$$

$$-\quad 7\frac{7}{18}$$

$$\rule{2cm}{0.4pt}$$

$$10\frac{11}{18}$$

8. $102 - 75\frac{11}{20}$

$$102 \to 101\frac{20}{20}$$

$$-\quad 75\frac{11}{20}$$

$$\rule{2cm}{0.4pt}$$

$$26\frac{9}{20}$$

- 63 -

Exercise #12. Subtract the following mixed numbers.

1. $7\frac{5}{8} - 5\frac{3}{4}$

$\overset{6}{\cancel{7}}\frac{5}{8} = \frac{5}{8} + \frac{8}{8} = \frac{13}{8}$

$5\frac{3}{4} = \frac{6}{8}$

$1\frac{7}{8}$

2. $13\frac{17}{34} - 10\frac{11}{17}$

$\overset{12}{\cancel{13}}\frac{17}{34} = \frac{17}{34} + \frac{34}{34} = \frac{51}{34}$

$10\frac{11}{17} = \frac{22}{34}$

$2\frac{29}{34}$

3. $12\frac{1}{4} - 11\frac{1}{3}$

$\overset{11}{\cancel{12}}\frac{1}{4} = \frac{3}{12} + \frac{12}{12} = \frac{15}{12}$

$11\frac{1}{3} = \frac{4}{12}$

$\frac{11}{12}$

4. $8\frac{3}{5} - 7\frac{3}{4}$

$\overset{7}{\cancel{8}}\frac{3}{5} = \frac{12}{20} + \frac{20}{20} = \frac{32}{20}$

$7\frac{3}{4} = \frac{15}{20}$

$\frac{17}{20}$

5. $5\frac{1}{6} - 2\frac{11}{12}$

$\overset{4}{\cancel{5}}\frac{1}{6} = \frac{2}{12} + \frac{12}{12} = \frac{14}{12}$

$2\frac{11}{12} = \frac{11}{12}$

$2\frac{3}{12} \rightarrow 2\frac{1}{4}$

6. $16\frac{5}{6} - 9\frac{9}{10}$

$\overset{15}{\cancel{16}}\frac{5}{6} = \frac{25}{30} + \frac{30}{30} = \frac{55}{30}$

$9\frac{9}{10} = \frac{27}{30}$

$6\frac{28}{30} \rightarrow 6\frac{14}{15}$

- 64 -

Exercise #12. Subtract the following fractions. Continued…

7. $21\frac{2}{3} - 14\frac{7}{8}$

$$\overset{20}{\cancel{21}}\frac{2}{3} = \frac{16}{24} + \frac{24}{24} = \frac{40}{24}$$

$$14\frac{7}{8} = \frac{21}{24}$$

————   ————————

$$6\frac{19}{24}$$

8. $48\frac{2}{7} - 22\frac{13}{35}$

$$\overset{47}{\cancel{48}}\frac{2}{7} = \frac{10}{35} + \frac{35}{35} = \frac{45}{35}$$

$$22\frac{13}{35} = \frac{13}{35}$$

————   ————————

$$25\frac{32}{35}$$

Multiplication

of

Fractions

and

Mixed Numbers

Overview of Multiplying Fractions

Example: Multiply $\dfrac{2}{3} \cdot \dfrac{3}{7}$

Multiply: $\dfrac{2}{3} \cdot \dfrac{3}{7}$

Step 1. Multiply the numerators.

Step 2. Then multiply the denominators.

$$\frac{2}{3} \cdot \frac{3}{7} = \frac{2 \cdot 3}{3 \cdot 7} = \frac{6}{21}$$

Step 3. Simplify result, if possible.

3 is the greatest common factor of 6 and 21. $3 \times 2 = 6$ $3 \times 7 = 21$

$$\frac{6 \div 3}{21 \div 3} \Rightarrow \frac{2}{7}$$

Exercise #1. Multiply the following fractions.

1. $\dfrac{1}{2} \cdot \dfrac{3}{5}$	2. $\dfrac{5}{8} \cdot \dfrac{2}{7}$
3. $\dfrac{3}{5} \cdot \dfrac{4}{7}$	4. $\dfrac{9}{11} \cdot \dfrac{6}{7}$

- 68 -

Session Comprehension

Date: _____

1. List steps or definitions that you have difficulty in comprehending.

2. Explain what is most important to understand when multiplying fractions.

3. Explain in your own words when it is necessary to simplify an answer or result after the multiplication of fractions.

4.

Illustrate the multiplication of the following fractions.	Explain each step of the process.
$$\frac{3}{5} \cdot \frac{4}{9}$$	

- 69 -

Is <u>Cross</u> <u>Multiplication</u> and <u>Cross</u> <u>Cancellation</u> the same?

No! **Cross Multiplication** **requires an equal symbol and the multiplication of diagonal numerals cross from the equal symbol.** ex. $\dfrac{2}{3} \times \dfrac{3}{4} \Rightarrow 2 \cdot 4 = 3 \cdot 3 \Rightarrow 8 = 9$

Cross Cancellation **is to simplify the diagonal numerals of two fractions being multiplied.**

ex. $\dfrac{2}{3} \cdot \dfrac{3}{4} \Rightarrow \dfrac{\overset{1}{2}}{\underset{1}{3}} \cdot \dfrac{\overset{3}{3}}{\underset{2}{4}}^{1} \Rightarrow \dfrac{1}{1} \cdot \dfrac{1}{2} \Rightarrow \dfrac{1}{2}$

Example (1) **Multiply:** $\dfrac{4}{5} \cdot \dfrac{15}{16}$

Step 1. Cross cancel the left diagonal. **4** is a factor of **4** and **16**.

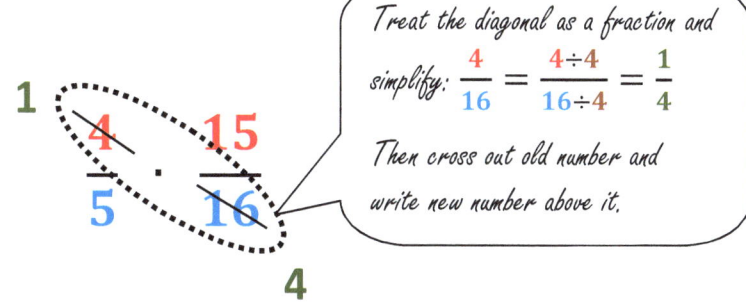

Treat the diagonal as a fraction and simplify: $\dfrac{4}{16} = \dfrac{4 \div 4}{16 \div 4} = \dfrac{1}{4}$

Then cross out old number and write new number above it.

Step 2. Cross cancel the left diagonal. **5** is a factor of **15** and **5**.

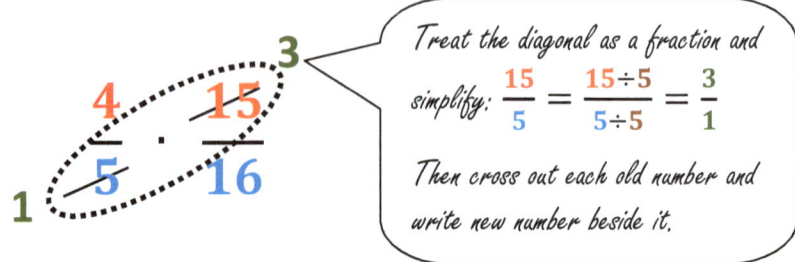

$$\frac{4}{5} \cdot \frac{\cancel{15}^{3}}{16}$$
$$1$$

Treat the diagonal as a fraction and simplify: $\dfrac{15}{5} = \dfrac{15 \div 5}{5 \div 5} = \dfrac{3}{1}$

Then cross out each old number and write new number beside it.

Step 3. Multiply numerators, then multiply denominators.

$$\frac{\cancel{4}^{1} \cdot \cancel{15}^{3}}{\cancel{5}_{1} \cdot \cancel{16}_{4}} = \frac{1 \cdot 3}{1 \cdot 4} = \boxed{\frac{3}{4}}$$

Step 3. Simplify result, if possible.

Note: Cross Cancellation does not guarantee that the result is fully simplified.

WARNING : Only diagonals with a common factor can be cross cancelled!!!

Example (2) **Multiply:** $\dfrac{7}{12} \cdot \dfrac{4}{9}$

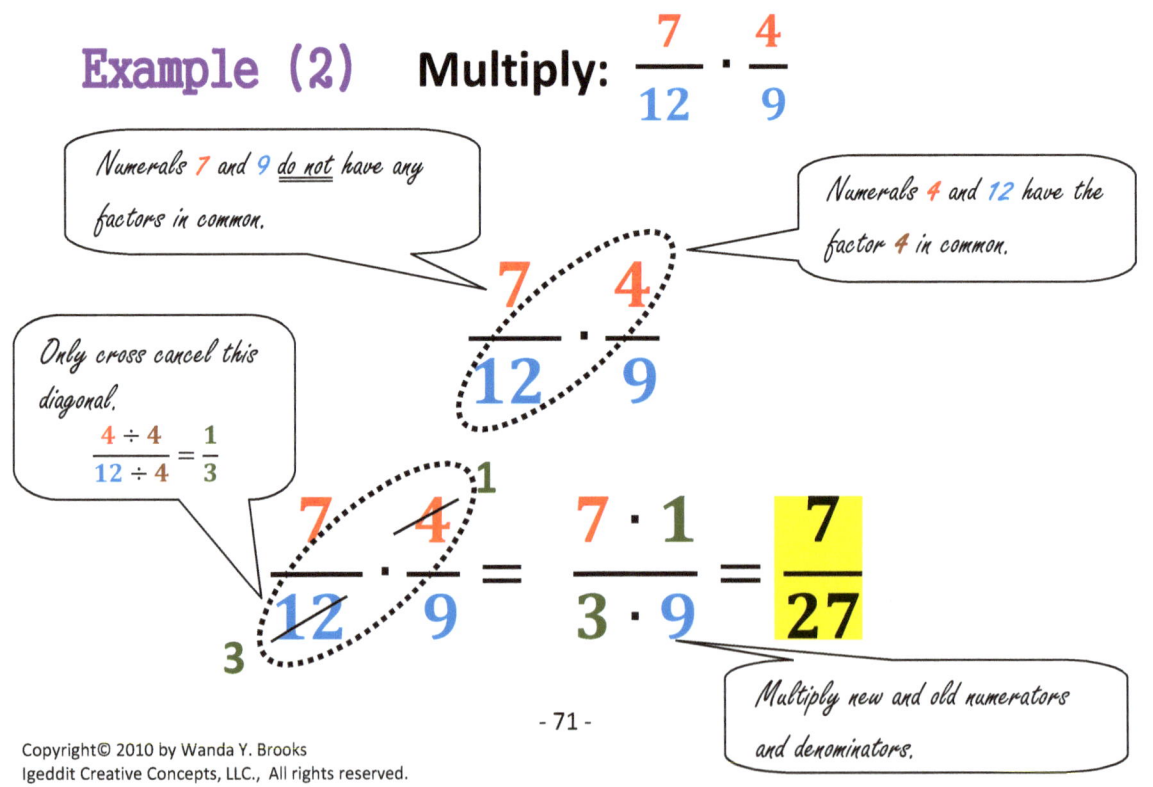

Numerals 7 and 9 <u>do not</u> have any factors in common.

Numerals 4 and 12 have the factor 4 in common.

Only cross cancel this diagonal. $\dfrac{4 \div 4}{12 \div 4} = \dfrac{1}{3}$

$$\frac{7}{\cancel{12}_{3}} \cdot \frac{\cancel{4}^{1}}{9} = \frac{7 \cdot 1}{3 \cdot 9} = \boxed{\frac{7}{27}}$$

Multiply new and old numerators and denominators.

Session Comprehension

Date: _____

1. List steps or definitions that you have difficulty in comprehending.

2. Explain what is most important to understand when cross cancelling fractions.

3. Explain in your own words how the process of cross cancellation is similar to the simplification or reduction of fractions.

4.

Illustrate the multiplication of the following fractions using the process of cross cancellation. $$\frac{5}{18} \cdot \frac{6}{25}$$	Explain each step of the process.
	⇨
	⇨

Exercise #2. Multiply the following fractions.

1. $\dfrac{2}{9} \cdot \dfrac{3}{8}$	2. $\dfrac{5}{12} \cdot \dfrac{3}{4}$
3. $\dfrac{12}{13} \cdot \dfrac{19}{24}$	4. $\dfrac{3}{5} \cdot \dfrac{10}{21}$
5. $\dfrac{4}{3} \cdot \dfrac{15}{16}$	6. $\dfrac{21}{2} \cdot \dfrac{1}{7}$
7. $\dfrac{7}{8} \cdot \dfrac{4}{7}$	8. $\dfrac{8}{15} \cdot \dfrac{5}{24}$

Multiplying Mixed Numbers

Example: Multiply $4\frac{2}{7} \cdot 3\frac{1}{2}$

How do we multiply mixed numbers?

Step 1. Change the mixed number(s) into improper fractions.
(see Retro Lesson C, page 8)

$$4\frac{2}{7} \cdot 3\frac{1}{2} \;\; becomes \;\; \frac{30}{7} \cdot \frac{7}{2}$$

Step 2. Proceed to cross cancel (if possible), then multiply the numerators and denominators.

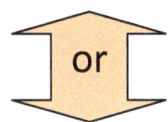

Directly multiply numerators and denominators without cross cancellation.

Cross Cancellation

$$\frac{\overset{15}{\cancel{30}}}{\underset{1}{\cancel{7}}} \cdot \frac{\overset{1}{\cancel{7}}}{\underset{1}{\cancel{2}}} = \frac{15 \cdot 1}{1 \cdot 1} = \frac{15}{1} \;\; or \;\; 15$$

$$\frac{30}{7} \cdot \frac{7}{2} = \frac{30 \cdot 7}{7 \cdot 2} = \frac{210}{14}$$

Basic Multiplication

Step 3. Simplify result (if possible).

$$\frac{210}{14} = \frac{210 \div 14}{14 \div 14} = \frac{15}{1} \;\; or \;\; 15$$

- 74 -

Multiplying a Fraction or Mixed Numbers, and a Whole Number

Example: Multiply $3\frac{1}{5} \cdot 10$

How do we multiply a mixed number and a Whole Number?

Example (1)

Step 1a. Change the <u>mixed number</u> into <u>an improper fraction</u>.

Step 1b. Change <u>the whole number</u> into <u>an improper fraction</u> by placing the **whole number** over **1**.

$$3\frac{1}{5} \cdot 10 \ \textit{becomes} \ \frac{16}{5} \cdot \frac{10}{1}$$

Step 2. Proceed to cross cancel and/or multiply.

Cross Cancellation

$$\frac{16}{\underset{1}{5}} \cdot \frac{\overset{2}{10}}{1} = \frac{16 \cdot 2}{1 \cdot 1} = \frac{32}{1} \ \textit{or} \ 32$$

Step 3. Simplify result (if possible).

Example (2)   Multiply: $\frac{2}{5} \cdot 3$

$$\frac{2}{5} \cdot 3 \ \textit{becomes} \ \frac{2}{5} \cdot \frac{3}{1} = \frac{6}{5} \ \textit{or} \ 1\frac{1}{5}$$

- 75 -

Session Comprehension

1. List steps or definitions that you have difficulty in comprehending.

2. Explain what is most important to understand when multiplying mixed numbers.

3. Explain in your own words the process of changing a mixed number to an improper fraction and the process of changing a whole number into an improper fraction.

4.

Illustrate the multiplication of the following mixed numbers, whole number and fractions.	Explain each step of the process.
$4\frac{1}{2} \cdot 8$	
$3\frac{3}{4} \cdot 2\frac{2}{3}$	

Exercise #3. Multiply the following mixed numbers.

1. $2\frac{2}{5} \cdot 6\frac{2}{3}$	2. $2\frac{1}{3} \cdot \frac{1}{7}$
3. $1\frac{3}{4} \cdot 8$	4. $\frac{1}{4} \cdot 3\frac{1}{5}$
5. $4\frac{1}{2} \cdot 5\frac{1}{3}$	6. $7 \cdot 5\frac{2}{7}$
7. $2\frac{1}{3} \cdot 1\frac{1}{5}$	8. $6\frac{1}{8} \cdot 4\frac{2}{7}$

- 77 -

Division
of
Fractions
and
Mixed Numbers

Division of Fractions

Example: Multiply $\dfrac{5}{12} \div \dfrac{5}{8}$

How do we divide fractions?

Step 1. Implement the **Keep** **Change** **Flip** technique.

Keep	Change	Flip
Keep the first fraction the same	Change division to multiplication	Change the second fraction to its **reciprocal** or flip
$\dfrac{5}{12}$	$\div$	$\dfrac{5}{8}$
$\dfrac{5}{12}$	$\times$	$\dfrac{8}{5}$

Step 2. Proceed to use the steps for multiplying fractions.

Using Basic Multiplication → $\dfrac{5}{12} \times \dfrac{8}{5} = \dfrac{40}{60} = \dfrac{2}{3}$

Using Cross Multiplication

5 is a factor of **5** and **5**

4 is a factor of **12** and **8**

$$\dfrac{\overset{1}{\cancel{5}}}{\underset{3}{\cancel{12}}} \times \dfrac{\overset{2}{\cancel{8}}}{\underset{1}{\cancel{5}}} = \dfrac{2}{3}$$

Step 3. Simplify result, if possible.

- 80 -

Example (1)

Divide: $\dfrac{10}{3} \div \dfrac{5}{9}$

$$\dfrac{10}{3} \div \dfrac{5}{9} \Rightarrow \dfrac{\overset{2}{\cancel{10}}}{\underset{1}{\cancel{3}}} \cdot \dfrac{\overset{3}{\cancel{9}}}{\underset{1}{\cancel{5}}} = \dfrac{6}{1} \; or \; 6$$

Or

$$\dfrac{10}{3} \div \dfrac{5}{9} \Rightarrow \dfrac{10}{3} \cdot \dfrac{9}{5} = \dfrac{90}{15} \; or \; 6$$

Example (2)

Divide: $\dfrac{16}{21} \div \dfrac{2}{35}$

$$\dfrac{16}{21} \div \dfrac{8}{35} \Rightarrow \dfrac{\overset{2}{\cancel{16}}}{\underset{3}{\cancel{21}}} \cdot \dfrac{\overset{5}{\cancel{35}}}{\underset{1}{\cancel{8}}} = \dfrac{10}{3} \; or \; 3\dfrac{1}{3}$$

Session Comprehension

Date: _____

1. List steps or definitions that you have difficulty in comprehending.

2. Explain what is most important to understand when dividing fractions.

3. Explain in your own words the keep change flip process and the similarities between the multiplication of fractions and the division of fractions.

4.

Illustrate the division of the following fractions.	Explain each step of the process.
$$\frac{9}{4} \div \frac{3}{8}$$	

Exercise #4. Divide the following fractions.

1. $\dfrac{1}{2} \div \dfrac{3}{5}$	2. $\dfrac{1}{4} \div \dfrac{7}{8}$
3. $\dfrac{4}{9} \div \dfrac{16}{15}$	4. $\dfrac{3}{8} \div \dfrac{9}{2}$
5. $\dfrac{2}{7} \div \dfrac{8}{21}$	6. $\dfrac{3}{7} \div \dfrac{12}{5}$
7. $\dfrac{3}{4} \div \dfrac{9}{16}$	8. $\dfrac{9}{4} \div \dfrac{3}{20}$

- 83 -

Dividing Mixed Numbers

Example: Divide $\quad 2\frac{2}{3} \div 1\frac{3}{5}$

How do we divide mixed numbers?

Step 1. Change the mixed number(s) into improper fractions.
(see Retro Lesson C, page 8)

$$2\frac{2}{3} \div 1\frac{3}{5} \quad becomes \quad \frac{8}{3} \div \frac{8}{5}$$

Step 2. Proceed to do the **Keep Change Flip**.

Warning: When dividing mixed numbers, never skip to step 2 before changing the mixed numbers to improper fractions.

$$\frac{8}{3} \div \frac{8}{5} \quad becomes \quad \frac{8}{3} \cdot \frac{5}{8}$$

Step 3. Then multiply or cross cancel and multiply.

Using Cross Cancellation

$$\frac{\overset{1}{\cancel{8}}}{3} \cdot \frac{5}{\underset{1}{\cancel{8}}} = \frac{5}{3} \ or \ 1\frac{2}{3}$$

Step 4. Simplify result (if possible).

Divide 40 by 8 and divide 24 by 8.

Using Basic Multiplication

$$\frac{8}{3} \cdot \frac{5}{8} = \frac{40}{24} = \frac{5}{3} \ or \ 1\frac{2}{3}$$

- 84 -

Dividing a Mixed Number and a Whole number

Example: Divide $\quad 4\dfrac{1}{2} \div 8$

How do we divide a mixed numbers and a whole number?

Example (1)

Step 1a. Change the <u>mixed number</u> into <u>an improper fraction</u>.

Step 1b. Change <u>the whole number</u> into <u>an improper fraction</u> by placing the **whole number** over **1**.

$$4\dfrac{1}{2} \div 8 \;\; becomes \;\; \dfrac{9}{2} \div \dfrac{8}{1}$$

Step 2. Proceed to do the **Keep Change Flip**, then multiply.

Note: Again, when dividing mixed numbers, never skip to step 2 before changing mixed numbers to improper fractions.

$$\dfrac{9}{2} \div \dfrac{8}{1} \;\; becomes \;\; \dfrac{9}{2} \cdot \dfrac{1}{8} = \dfrac{9}{16}$$

Step 3. Simplify, if possible.

Using Basic Multiplication

Example (2)

Divide: $\quad 10 \div \dfrac{15}{17}$

$$\dfrac{10}{1} \div \dfrac{15}{17} \Rightarrow \dfrac{\overset{2}{\cancel{10}}}{1} \cdot \dfrac{17}{\underset{3}{\cancel{15}}} = \dfrac{34}{3} \;\; or \;\; 11\dfrac{1}{3}$$

Session Comprehension

Date: _____

1. List steps or definitions that you have difficulty in comprehending.

2. Explain what is most important to understand when dividing mixed numbers.

3. Explain in your own words the similarities between the multiplication of mixed numbers and the division of mixed numbers.

4.

Illustrate the division of the following mixed numbers.	Explain each step of the process.
$$3\frac{1}{3} \div 2\frac{2}{9}$$	

Exercise #5. Divide the following mixed numbers.

1. $\dfrac{5}{18} \div 3$	2. $12 \div \dfrac{2}{3}$
3. $2\dfrac{1}{2} \div 4$	4. $7 \div 5\dfrac{1}{4}$
5. $6\dfrac{2}{3} \div 3\dfrac{1}{3}$	6. $11\dfrac{1}{3} \div 8\dfrac{3}{4}$
7. $\dfrac{1}{2} \div 2\dfrac{1}{4}$	8. $2\dfrac{2}{5} \div 1\dfrac{3}{5}$

Answer

B

Exercise #1. Multiply the following fractions.

1. $\dfrac{1}{2} \cdot \dfrac{3}{5}$ $\dfrac{1}{2} \cdot \dfrac{3}{5} = \dfrac{1 \cdot 3}{2 \cdot 5} = \dfrac{3}{10}$	2. $\dfrac{5}{8} \cdot \dfrac{2}{7}$ $\dfrac{5}{8} \cdot \dfrac{2}{7} = \dfrac{5 \cdot 2}{8 \cdot 7} = \dfrac{10 \div 2}{56 \div 2} = \dfrac{5}{28}$
3. $\dfrac{3}{5} \cdot \dfrac{4}{7}$ $\dfrac{3}{5} \cdot \dfrac{4}{7} = \dfrac{3 \cdot 4}{5 \cdot 7} = \dfrac{12}{35}$	4. $\dfrac{9}{11} \cdot \dfrac{6}{7}$ $\dfrac{9}{11} \cdot \dfrac{6}{7} = \dfrac{9 \cdot 6}{11 \cdot 7} = \dfrac{54}{77}$

Exercise #2. Multiply the following fractions.

1. $\dfrac{2}{9} \cdot \dfrac{3}{8}$

$$\dfrac{\overset{1}{\cancel{2}}}{\underset{3}{\cancel{9}}} \cdot \dfrac{\overset{1}{\cancel{3}}}{\underset{4}{\cancel{8}}} = \dfrac{1 \cdot 1}{3 \cdot 4} = \boxed{\dfrac{1}{12}}$$

2. $\dfrac{5}{12} \cdot \dfrac{3}{4}$

$$\dfrac{5}{\underset{4}{\cancel{12}}} \cdot \dfrac{\overset{1}{\cancel{3}}}{4} = \boxed{\dfrac{5}{16}}$$

3. $\dfrac{12}{13} \cdot \dfrac{19}{24}$

$$\dfrac{\overset{1}{\cancel{12}}}{13} \cdot \dfrac{19}{\underset{2}{\cancel{24}}} = \boxed{\dfrac{19}{26}}$$

4. $\dfrac{3}{5} \cdot \dfrac{10}{21}$

$$\dfrac{\overset{1}{\cancel{3}}}{\underset{1}{\cancel{5}}} \cdot \dfrac{\overset{2}{\cancel{10}}}{\underset{7}{\cancel{21}}} = \boxed{\dfrac{2}{7}}$$

5. $\dfrac{4}{3} \cdot \dfrac{15}{16}$

$$\dfrac{\overset{1}{\cancel{4}}}{\underset{1}{\cancel{3}}} \cdot \dfrac{\overset{5}{\cancel{15}}}{\underset{4}{\cancel{16}}} = \boxed{\dfrac{5}{4}} \; or \; 1\dfrac{1}{4}$$

6. $\dfrac{21}{2} \cdot \dfrac{1}{7}$

$$\dfrac{\overset{3}{\cancel{21}}}{2} \cdot \dfrac{1}{\underset{1}{\cancel{7}}} = \boxed{\dfrac{3}{2}} \; or \; 1\dfrac{1}{2}$$

7. $\dfrac{7}{8} \cdot \dfrac{4}{7}$

$$\dfrac{\overset{1}{\cancel{7}}}{\underset{2}{\cancel{8}}} \cdot \dfrac{\overset{1}{\cancel{4}}}{\underset{1}{\cancel{7}}} = \boxed{\dfrac{1}{2}}$$

8. $\dfrac{8}{15} \cdot \dfrac{5}{24}$

$$\dfrac{\overset{1}{\cancel{8}}}{\underset{3}{\cancel{15}}} \cdot \dfrac{\overset{1}{\cancel{5}}}{\underset{3}{\cancel{24}}} = \boxed{\dfrac{1}{9}}$$

Exercise #3. Multiply the following mixed numbers.

1. $2\frac{2}{5} \cdot 6\frac{2}{3}$

$$\overset{4}{\underset{1}{\cancel{\frac{12}{5}}}} \cdot \overset{4}{\underset{1}{\cancel{\frac{20}{3}}}} = \frac{16}{1} \text{ or } 16$$

2. $2\frac{1}{3} \cdot \frac{1}{7}$

$$\overset{1}{\cancel{\frac{7}{3}}} \cdot \underset{1}{\cancel{\frac{1}{7}}} = \frac{1}{3}$$

3. $1\frac{3}{4} \cdot 8$

$$\underset{1}{\frac{7}{4}} \cdot \overset{2}{\cancel{\frac{8}{1}}} = \frac{14}{1} \text{ or } 14$$

4. $\frac{1}{4} \cdot 3\frac{1}{5}$

$$\underset{1}{\frac{1}{4}} \cdot \overset{4}{\cancel{\frac{16}{5}}} = \frac{4}{5}$$

5. $4\frac{1}{2} \cdot 5\frac{1}{3}$

$$\overset{3}{\underset{1}{\cancel{\frac{9}{2}}}} \cdot \overset{8}{\underset{1}{\cancel{\frac{16}{3}}}} = \frac{24}{1} \text{ or } 24$$

6. $7 \cdot 5\frac{2}{7}$

$$\overset{1}{\cancel{\frac{7}{1}}} \cdot \underset{1}{\cancel{\frac{37}{7}}} = \frac{37}{1} \text{ or } 37$$

7. $2\frac{1}{3} \cdot 1\frac{1}{5}$

$$\underset{1}{\cancel{\frac{7}{3}}} \cdot \overset{2}{\cancel{\frac{6}{5}}} = \frac{14}{5} \text{ or } 2\frac{4}{5}$$

8. $6\frac{1}{8} \cdot 4\frac{2}{7}$

$$\overset{7}{\underset{4}{\cancel{\frac{49}{8}}}} \cdot \overset{15}{\underset{1}{\cancel{\frac{30}{7}}}} = \frac{105}{4} \text{ or } 26\frac{1}{4}$$

- 92 -

Exercise #4. Divide the following fractions.

1. $\dfrac{1}{2} \div \dfrac{3}{5}$

$$\dfrac{1}{2} \div \dfrac{3}{5} \Rightarrow \dfrac{1}{2} \cdot \dfrac{5}{3} = \dfrac{5}{6}$$

2. $\dfrac{1}{4} \div \dfrac{7}{8}$

$$\dfrac{1}{4} \div \dfrac{7}{8} \Rightarrow \dfrac{1}{\underset{1}{4}} \cdot \dfrac{\overset{2}{8}}{7} = \dfrac{2}{7}$$

3. $\dfrac{4}{9} \div \dfrac{16}{15}$

$$\dfrac{4}{9} \div \dfrac{16}{15} \Rightarrow \dfrac{\overset{1}{4}}{\underset{3}{9}} \cdot \dfrac{\overset{5}{15}}{\underset{4}{16}} = \dfrac{5}{12}$$

4. $\dfrac{3}{8} \div \dfrac{9}{2}$

$$\dfrac{3}{8} \div \dfrac{9}{2} \Rightarrow \dfrac{\overset{1}{3}}{\underset{4}{8}} \cdot \dfrac{\overset{1}{2}}{\underset{3}{9}} = \dfrac{1}{12}$$

5. $\dfrac{2}{7} \div \dfrac{8}{21}$

$$\dfrac{2}{7} \div \dfrac{8}{21} \Rightarrow \dfrac{\overset{1}{2}}{\underset{1}{7}} \cdot \dfrac{\overset{3}{21}}{\underset{4}{8}} = \dfrac{3}{4}$$

6. $\dfrac{3}{7} \div \dfrac{12}{5}$

$$\dfrac{3}{7} \div \dfrac{12}{5} \Rightarrow \dfrac{\overset{1}{3}}{7} \cdot \dfrac{5}{\underset{4}{12}} = \dfrac{5}{28}$$

7. $\dfrac{3}{4} \div \dfrac{9}{16}$

$$\dfrac{3}{4} \div \dfrac{9}{16} \Rightarrow \dfrac{\overset{1}{3}}{\underset{1}{4}} \cdot \dfrac{\overset{4}{16}}{\underset{3}{9}} = \dfrac{4}{3} \; or \; 1\dfrac{1}{3}$$

8. $\dfrac{9}{4} \div \dfrac{3}{20}$

$$\dfrac{9}{4} \div \dfrac{3}{30} \Rightarrow \dfrac{\overset{3}{9}}{\underset{1}{4}} \cdot \dfrac{\overset{5}{20}}{\underset{1}{3}} = \dfrac{15}{1} \; or \; 15$$

- 93 -

Exercise #5. Divide the following mixed numbers.

1. $\dfrac{5}{18} \div 3$ $\dfrac{5}{18} \div 3 \Rightarrow \dfrac{5}{18} \div \dfrac{3}{1} \Rightarrow \dfrac{5}{18} \cdot \dfrac{1}{3} = \dfrac{5}{54}$	**2.** $12 \div \dfrac{2}{3}$ $12 \div \dfrac{2}{3} \Rightarrow \dfrac{12}{1} \div \dfrac{2}{3} \Rightarrow \dfrac{\overset{6}{\cancel{12}}}{1} \cdot \dfrac{3}{\underset{1}{\cancel{2}}} = \dfrac{18}{1} = 18$
3. $2\dfrac{1}{2} \div 4$ $2\dfrac{1}{2} \div 4 \Rightarrow \dfrac{5}{2} \div \dfrac{4}{1} \Rightarrow \dfrac{5}{2} \cdot \dfrac{1}{4} = \dfrac{5}{8}$	**4.** $7 \div 5\dfrac{1}{4}$ $7 \div 5\dfrac{1}{4} \Rightarrow \dfrac{7}{1} \div \dfrac{21}{4} \Rightarrow \dfrac{\overset{1}{\cancel{7}}}{1} \cdot \dfrac{4}{\underset{3}{\cancel{21}}} = \dfrac{4}{3}$
5. $6\dfrac{2}{3} \div 3\dfrac{1}{3}$ $6\dfrac{2}{3} \div 3\dfrac{1}{3} \Rightarrow \dfrac{20}{3} \div \dfrac{10}{3} \Rightarrow \dfrac{\overset{2}{\cancel{20}}}{3} \cdot \dfrac{3}{\underset{1}{\cancel{10}}} = 2$	**6.** $11\dfrac{1}{3} \div 8\dfrac{3}{4}$ $11\dfrac{1}{3} \div 8\dfrac{3}{4} \Rightarrow \dfrac{34}{3} \div \dfrac{35}{4} \Rightarrow \dfrac{34}{3} \cdot \dfrac{4}{35} = \dfrac{136}{105} \; or \; 1\dfrac{31}{105}$
7. $\dfrac{1}{2} \div 2\dfrac{1}{4}$ $\dfrac{1}{2} \div 2\dfrac{1}{4} \Rightarrow \dfrac{1}{2} \div \dfrac{9}{4} \Rightarrow \dfrac{1}{\underset{1}{\cancel{2}}} \cdot \dfrac{\overset{2}{\cancel{4}}}{9} = \dfrac{2}{9}$	**8.** $2\dfrac{2}{5} \div 1\dfrac{3}{5}$ $2\dfrac{2}{5} \div 1\dfrac{3}{5} \Rightarrow \dfrac{12}{5} \div \dfrac{8}{5} \Rightarrow \dfrac{\overset{3}{\cancel{12}}}{\underset{1}{\cancel{5}}} \cdot \dfrac{\cancel{5}}{\underset{2}{\cancel{8}}} = \dfrac{3}{2} \; or \; 1\dfrac{1}{2}$

- 94 -

BONUS

Comparing Fractions

Fractions must have a common denominator in order to determine the larger or smaller fraction.

Example:

Which symbol is correct, < or > , for the comparison of $\dfrac{3}{16}$ $\dfrac{5}{8}$?

Step 1. Find the least common denominator.

The L.C.D. of 16 and 8 is **16**.

Step 2. Create Equivalent fractions with the L.C.D.

$$\frac{3\cdot1}{16\cdot1} = \frac{3}{16} \quad \text{and} \quad \frac{5\cdot2}{8\cdot2} = \frac{10}{16}$$

Step 3. Then compare numerators.

3 is smaller than **10** , therefore $\dfrac{3}{16} < \dfrac{5}{8}$

Bonus exercise #1. Compare the fractions using the symbols <, > or = .

1. $\dfrac{2}{6}$	$\dfrac{3}{9}$		2. $\dfrac{15}{24}$	$\dfrac{11}{18}$	
3. $\dfrac{5}{12}$	$\dfrac{3}{7}$		4. $\dfrac{9}{24}$	$\dfrac{3}{8}$	
5. $\dfrac{2}{9}$	$\dfrac{4}{14}$		6. $\dfrac{14}{15}$	$\dfrac{4}{5}$	

- 96 -

BONUS

Comparing Fractions for Equivalency

Use **CROSS MULTIPLICATION** when determining if two fractions are equal.

Example: Is this statement true, $\dfrac{5}{9} = \dfrac{15}{27}$?

Step 1. Cross multiply the diagonals.

$$\dfrac{5}{9} \diagup\!\!\!\!\diagdown \dfrac{15}{27}$$

$$5 \cdot 27 = 15 \cdot 9$$
$$135 = 135$$

Step 3. Compare the sides.

    a.   If both sides are <mark>equal</mark>, then the fractions <mark>are equivalent</mark>.

    b.   If both sides are <mark>not equal</mark>, then the fractions are <mark>not equivalent</mark>.

Therefore, fractions $\dfrac{5}{9}$ and $\dfrac{15}{27}$ are equivalent.

Bonus exercise #2. Determine whether the given fractions are equivalent.

1. $\dfrac{2}{6} = \dfrac{3}{9}$	2. $\dfrac{15}{24} = \dfrac{11}{18}$
3. $\dfrac{5}{12} = \dfrac{3}{7}$	4. $\dfrac{9}{24} = \dfrac{3}{8}$
5. $\dfrac{6}{9} = \dfrac{10}{15}$	6. $\dfrac{6}{15} = \dfrac{2}{5}$

Bonus

Answers

Bonus exercise #1. Compare the fractions using the symbols <, > or = .

1. $\dfrac{2}{6} = \dfrac{3}{9}$	2. $\dfrac{15}{24} > \dfrac{11}{18}$
3. $\dfrac{5}{12} < \dfrac{3}{7}$	4. $\dfrac{9}{24} = \dfrac{3}{8}$
5. $\dfrac{2}{9} < \dfrac{4}{14}$	6. $\dfrac{14}{15} > \dfrac{4}{5}$

Bonus exercise #2. Determine whether the given fractions are equivalent.

1. $\dfrac{2}{6} = \dfrac{3}{9}$ equivalent	2. $\dfrac{15}{24} = \dfrac{11}{18}$ not equivalent
3. $\dfrac{5}{12} = \dfrac{3}{7}$ not equivalent	4. $\dfrac{9}{24} = \dfrac{3}{8}$ equivalent
5. $\dfrac{6}{9} = \dfrac{10}{15}$ equivalent	6. $\dfrac{6}{15} = \dfrac{2}{5}$ equivalent

www.ingramcontent.com/pod-product-compliance
Lightning Source LLC
Chambersburg PA
CBHW050728180526
45159CB00003B/1163